HOW BIG IS GLOBAL WARMING?

How big is Global Warming?

KIWI AARDVARK

Kiwi Aardvark

Contents

HOW BIG IS GLOBAL WARMING?

LESS THAN THE AVERAGE GLOBAL WARMING

MORE THAN THE AVERAGE GLOBAL WARMING

ACTUAL AVERAGE TEMPERATURES

ACTUAL WINTER TEMPERATURES

WHAT CAN WE COMPARE GLOBAL WARMING TO?

SEASONAL WARMING (WINTER TO SUMMER)

THE ACTUAL TEMPERATURE IN DIFFERENT CITIES

THE DISTANCE TO MOVE TO REVERSE GLOBAL WARMING

SUMMARY

How big is Global Warming?

Many people are very worried about global warming. They think that global warming is a big threat to the planet, including all humans, animals, and plants. But is this level of worry justified? This book will help you to answer that question.

The average amount of global warming for the entire Earth is about 1.1 degrees Celsius of warming since 1880. But this amount of warming applies to the average of both the Oceans and the Land. The average amount of warming for just the Land (where most people live) is currently about 1.8 degrees Celsius, since 1880.

However, the Northern Hemisphere Land has warmed more than the Southern Hemisphere Land. The Northern Hemisphere Land has warmed by about 2.0 degrees Celsius since 1880, while the Southern Hemisphere Land has warmed by about 1.5 degrees Celsius since 1880.

This book divides the Earth into 14 regions, and then looks at how much warming has occurred in each region. Seven regions have warmed by less than the average, and seven regions have warmed by more than the average. What about where you live? Have you experienced less or more warming than the average amount?

Here is a map showing the 14 regions.

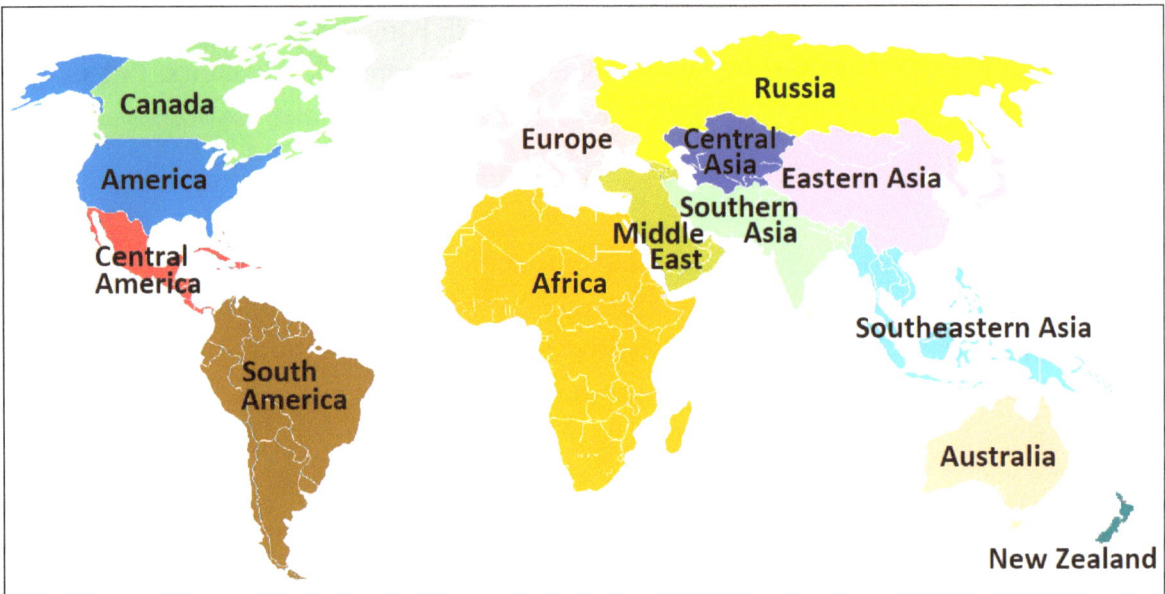

Figure 1. The 14 regions

The 14 regions looked at are:

- Southeast Asia
- Australia
- America
- Africa
- New Zealand
- Central America
- South Asia
- East Asia
- South America
- Central Asia
- Middle East
- Europe
- Canada
- Russia

Less than the average Global Warming

The average amount of warming on just the Land (where most people live) is currently about 1.8 degrees Celsius, since 1880.

The regions which have warmed by less than the average amount of warming are:

Country/Region	Amount of warming since 1880 (degrees Celsius)
Southeast Asia	1.0
Australia	1.2
America	1.2
Africa	1.2
New Zealand	1.3
Central America	1.4
South Asia	1.6
Land Average	1.8

These regions don't need to worry about warming as much as the regions which have had more than the average amount of warming.

Southeast Asia has only had just over half of the average amount of warming.

More than the average Global Warming

The average amount of warming on just the Land (where most people live) is currently about 1.8 degrees Celsius, since 1880.

The regions which have warmed by more than the average amount of warming are:

Country/Region	Amount of warming since 1880 (degrees Celsius)
Land Average	1.8
East Asia	1.9
South America	2.1
Central Asia	2.2
Middle East	2.2
Europe	2.5
Canada	2.6
Russia	2.7

Russia, Canada, and Europe have had about 50% more warming than the average amount of warming.

Russia, Canada, and Europe have had about 2.5 times as much warming as Eastern Asia.

An important question to ask is, should the regions which have had more warming than average since 1880 be more worried about global warming.

And should the regions which have had less warming than average since 1880 be less worried about global warming.

These questions can not be answered by just looking at the temperature change since 1880. Because we also need to consider whether these regions were actually hot or cold regions to begin with. For example, Russia, which has had more warming since 1880 than any other region, is actually the coldest region. They will probably be happy to have some warming.

Actual average temperatures

This map shows that many of the regions with the most warming since 1880 (for example, East Asia, Central Asia, Europe, Canada, and Russia) are cold regions (they have cold actual average temperatures).

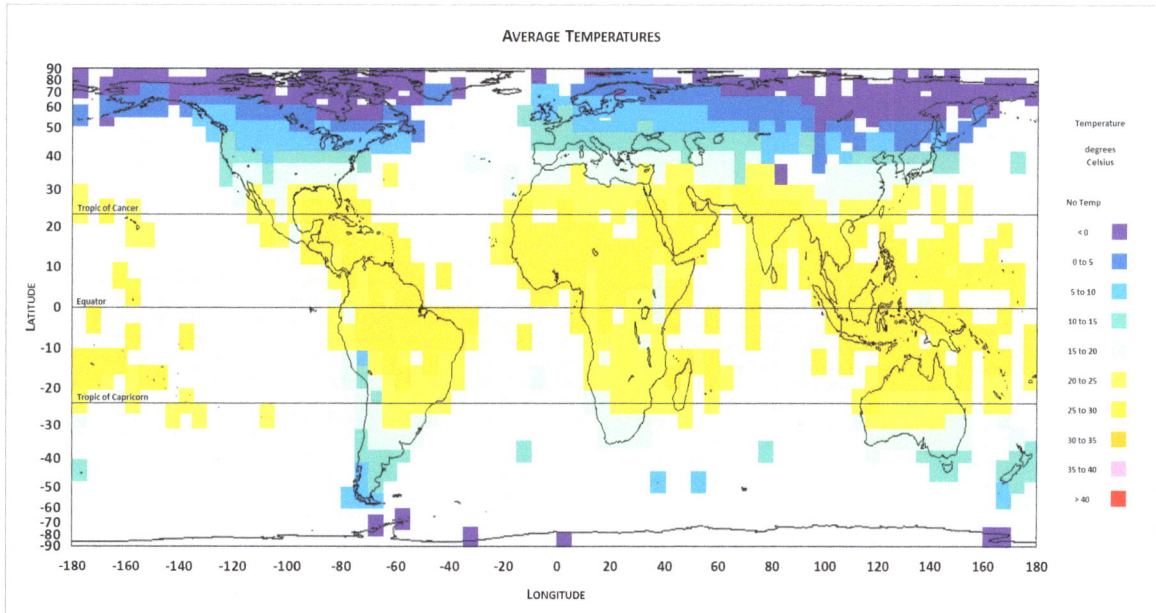

Figure 2. Actual average temperatures

Cold regions will probably welcome some global warming, because it will make them more like "average" countries.

Actual winter temperatures

This map shows that many of the regions with the most warming since 1880 (for example, East Asia, Central Asia, Europe, Canada, and Russia) have cold actual winter temperatures.

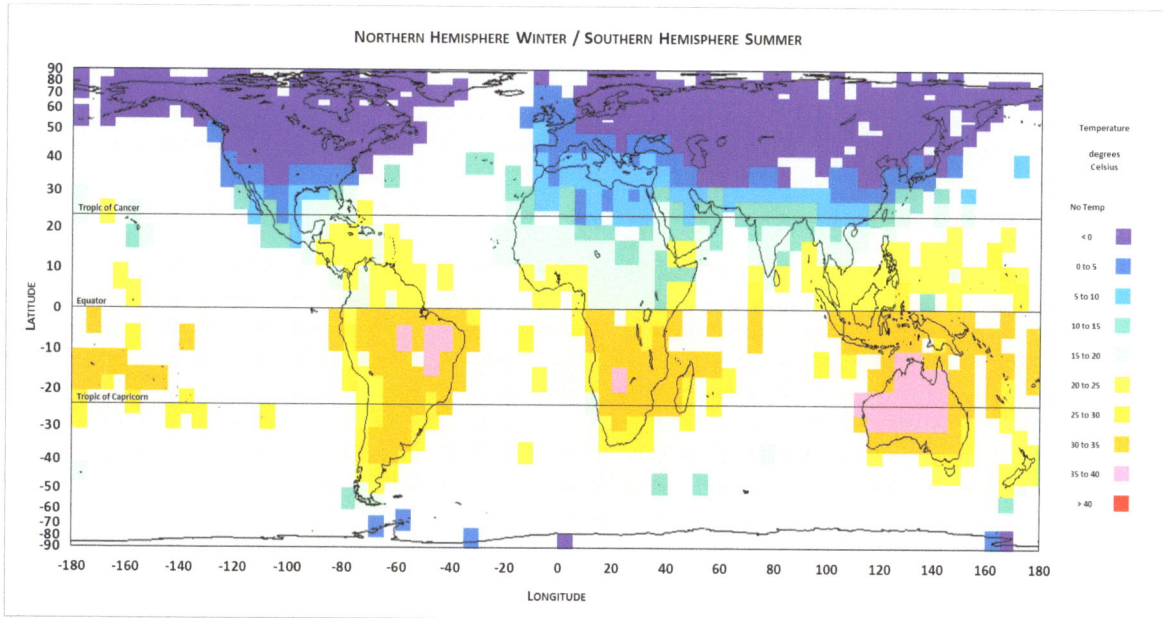

Figure 3. Actual winter temperatures

Most of the locations north of the Tropic of Cancer go below zero degrees Celsius every year in winter (they freeze, the purple colour on the map). Global warming will make their winters less severe.

Cold regions will probably welcome some global warming, because it will make them more like "average" countries.

What can we compare Global Warming to?

Is 1.0 degrees Celsius of warming bad?

Is 2.0 degrees Celsius of warming bad?

Is 3.0 degrees Celsius of warming bad?

Does it matter whether the region was hot or cold before the warming began?

It is difficult to know how bad global warming will be, because we don't have many other examples of warming to compare global warming to.

But there are other types of warming that we can compare global warming to. This book will look at 3 other examples of warming. These are

- seasonal warming (winter to summer)

- the actual temperature in different cities

- the distance to move to reverse global warming

Seasonal warming (winter to summer)

Seasonal warming is an example of warming that people experience every year. This is the warming that occurs when going from winter to summer, over a 6 month period. Most locations on the Earth warm by between 20 and 30 degrees Celsius when they change from winter to summer.

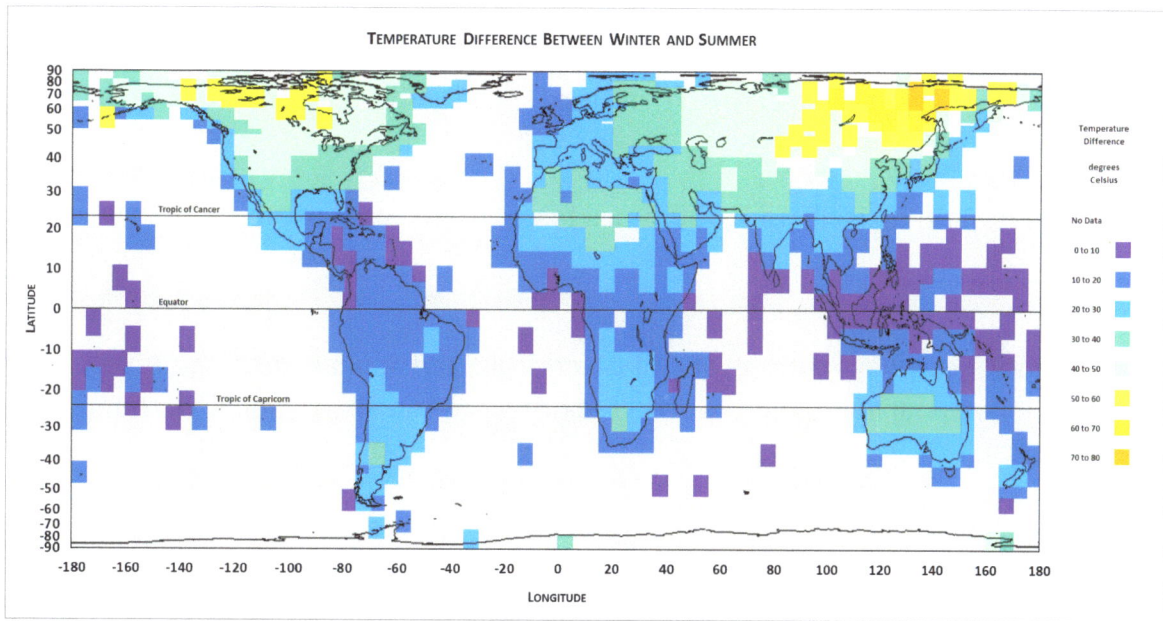

Figure 4. Winter/Summer temperature difference

Some locations even change by as much as 70 to 80 degrees Celsius when they change from winter to summer.

Most locations have a temperature increase of 20 to 30 degrees Celsius between winter and summer, and don't have any serious problems from this amount of warming. In fact, most people prefer summer (warmer temperatures) to winter (colder temperatures). But people are very worried about a small increase from global warming (another 0.5 or 1.0 degrees Celsius).

The seasonal warming of about 25 degrees Celsius happens over 6 months. So the speed is very fast, the equivalent of 5,000 degrees Celsius per century.

Global warming is only occurring at the speed of about 2 degrees Celsius per century.

So seasonal warming is about 2,500 times faster than global warming.

In summary, seasonal warming very fast and big, but global warming is very slow and small. But people are more worried about slow and small global warming. This is not logical.

Figure 5. Seasonal warming is like being chased by a cheetah

Figure 6. Global warming is like being chased by a snail

The actual temperature in different cities

How dangerous is it to increase your average temperature by 2 degrees Celsius?

How dangerous is it to increase your average temperature by 4 degrees Celsius?

How about 8 degrees Celsius?

Or how about 12 degrees Celsius?

Is it possible to survive if you increase your average temperature by 16 degrees Celsius?

Could anybody survive an increase in their average temperature by 20 degrees Celsius?

You might be surprised to learn that there are many people who increase their average temperature by more than 20 degrees Celsius, and live to tell the story. Many of these people are happy to have increased their average temperature by more than 20 degrees Celsius. They actually pay money to get that temperature increase.

Different cities in the world have different average temperatures.

Some cities have low average temperatures. Like Moscow, in Russia. Moscow's average temperature is about +4.0 degrees Celsius.

Some cities have high average temperatures. Like Bangkok, in Thailand. Bangkok's average temperature is about +28.0 degrees Celsius.

It is quite safe for a person who has lived in Moscow for their whole life, to hop on a plane and fly to Bangkok. Their average temperature will increase from +4.0 degrees Celsius to +28.0 degrees Celsius, within a day or two.

If they decide to live permanently in Bangkok, then their average temperature will increase permanently by 24.0 degrees Celsius. It may take them a little while to feel comfortable, but most people would survive their average temperature increasing by 24.0 degrees Celsius.

You can work out the temperature increase or decrease for many city moves, using the following average temperatures.

City	Country	Average Temp (degrees Celsius)
Moscow	Russia	+4.0
Helsinki	Finland	+5.0
Oslo	Norway	+6.0
Stockholm	Sweden	+6.0
Ottawa	Canada	+6.6
Kiev	Ukraine	+7.0
Berlin	Germany	+7.3
Bern	Switzerland	+7.9
Copenhagen	Denmark	+8.0
Warsaw	Poland	+8.0
Dublin	Ireland	+9.8
Amsterdam	Netherlands	+10.0
Ankara	Turkey	+10.0
London	England	+10.3
Beijing	China	+12.0
Paris	France	+12.4
Madrid	Spain	+14.1
Washington DC	America	+14.6
Rome	Italy	+15.5
Tokyo	Japan	+16.3

City	Country	Average Temp (degrees Celsius)
Sydney	Australia	+17.0
Mexica City	Mexica	+17.0
Tehran	Iran	+17.0
Cape Town	South Africa	+17.0
Buenos Aires	Argentina	+17.7
Kathmandu	Nepal	+18.0
Athens	Greece	+18.5
Nairobi	Kenya	+19.0
Lima	Peru	+20.0
Kampala	Uganda	+20.0
Brasilia	Brazil	+20.6
Cairo	Egypt	+21.0
Islamabad	Pakistan	+21.0
Baghdad	Iraq	+22.0
Hong Kong	China	+23.0
Delhi	India	+25.2
Jakarta	Indonesia	+27.0
Manila	Philippines	+27.0
Singapore	Singapore	+27.0
Bangkok	Thailand	+28.0

Millions of people live happily in these cities, at many different average temperatures. Is it logical to be very worried about another 0.5 or 1.0 degrees Celsius of global warming when you see the huge range of temperatures that people live at?

With international travel, and tourism, many people are constantly changing the average temperature that they live at, with little or no problems.

People who are worried about global warming may protest, "But what about animals and plants?"

There may be restrictions on which animals and plants are allowed into some countries. But this is not because the change in average temperature is dangerous for them. The reason is almost the opposite of this concern. Many animals and plants would do so well at the different average temperatures, that they could be a threat to native animals and plants.

To show how much average temperatures vary on the Earth, here is a graph which shows the Yearly Average Temperature versus Latitude for over 36,000 locations on the Earth. All of the locations are on Land.

There is not much land in the Southern Hemisphere between about -50 South and -70 South (Antarctica is not shown).

Compare this to the Northern Hemisphere where there is a lot of land from the Equator (0 N) to about 70 N (the Arctic is not shown).

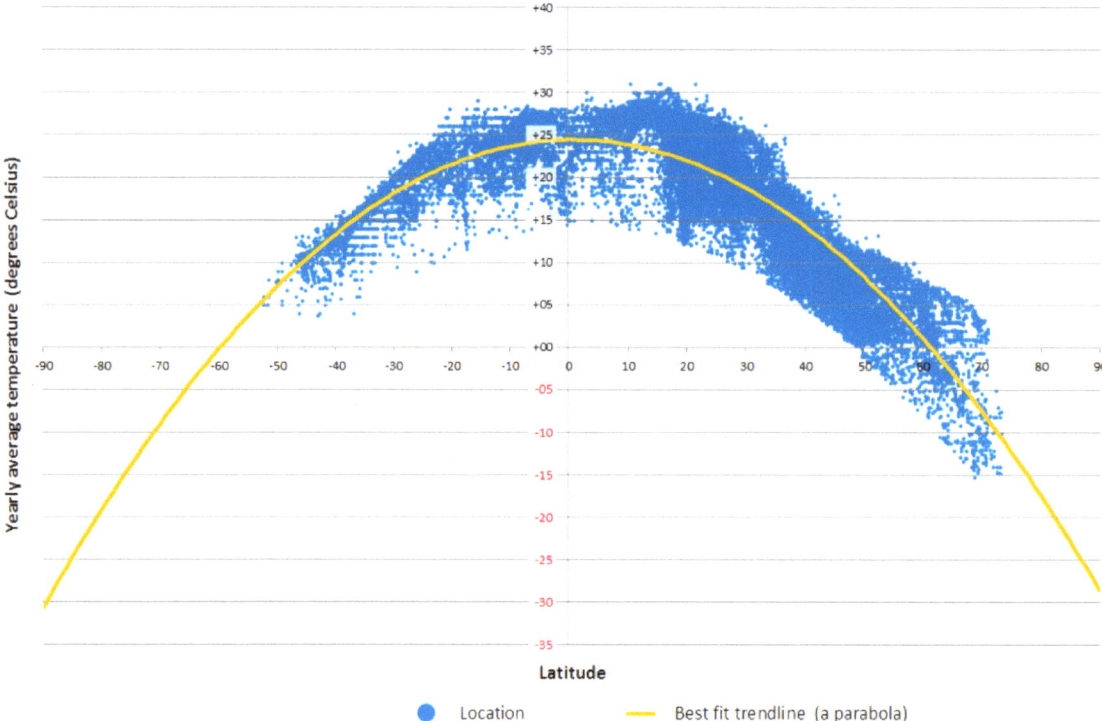

Yearly average temperature versus Latitude for over 36,000 locations

Yearly average temperature (degrees Celsius)

Latitude

● Location   — Best fit trendline (a parabola)

The distance to move to reverse global warming

There is a temperature gradient which goes from the Equator (at about +40 degrees Celsius) to the Poles (at about -40 degrees Celsius). The yearly average temperature varies with the Latitude.

So if you wanted to reverse the warming caused by global warming, one way that you could do this is to move towards the nearest Pole. But how far would you have to move?

It is important to realise that all of the locations that are at a particular latitude don't have exactly the same yearly average temperature. Latitude is the major factor that determines the yearly average temperature. But there are a number of other factors like

- elevation
- UHI (the Urban Heat Island Effect)
- the climate (e.g. if the climate is arid or dry)
- the proximity to the ocean
- the presence of ocean currents (e.g. the Gulf Stream)
- etc

All of these factors can also affect the yearly average temperature.

How far would you need to move, towards the nearest Pole (the North Pole, or the South Pole), to reverse the temperature increase from 1 additional degree Celsius of global warming?

We have seen how there is a temperature gradient, which goes from the equator (hottest), to the poles (coldest). So if you would like to live at a cooler temperature, all that you need to do is get into your car, and drive towards the nearest pole. But how far should you drive.

Let us assume that you live in location "A", which is at latitude "L". The distance that you have to drive depends upon the latitude "L". The nearer that you are to a Pole, the less distance that you have to drive for a given temperature change.

To put that another way, the nearer that you are to the equator, the bigger the distance that you have to drive for a given temperature change.

Look at the graph on the next page to see the distance (in kilometers), that you would need to move towards the nearest Pole, to reverse 1 degree Celsius of global warming.

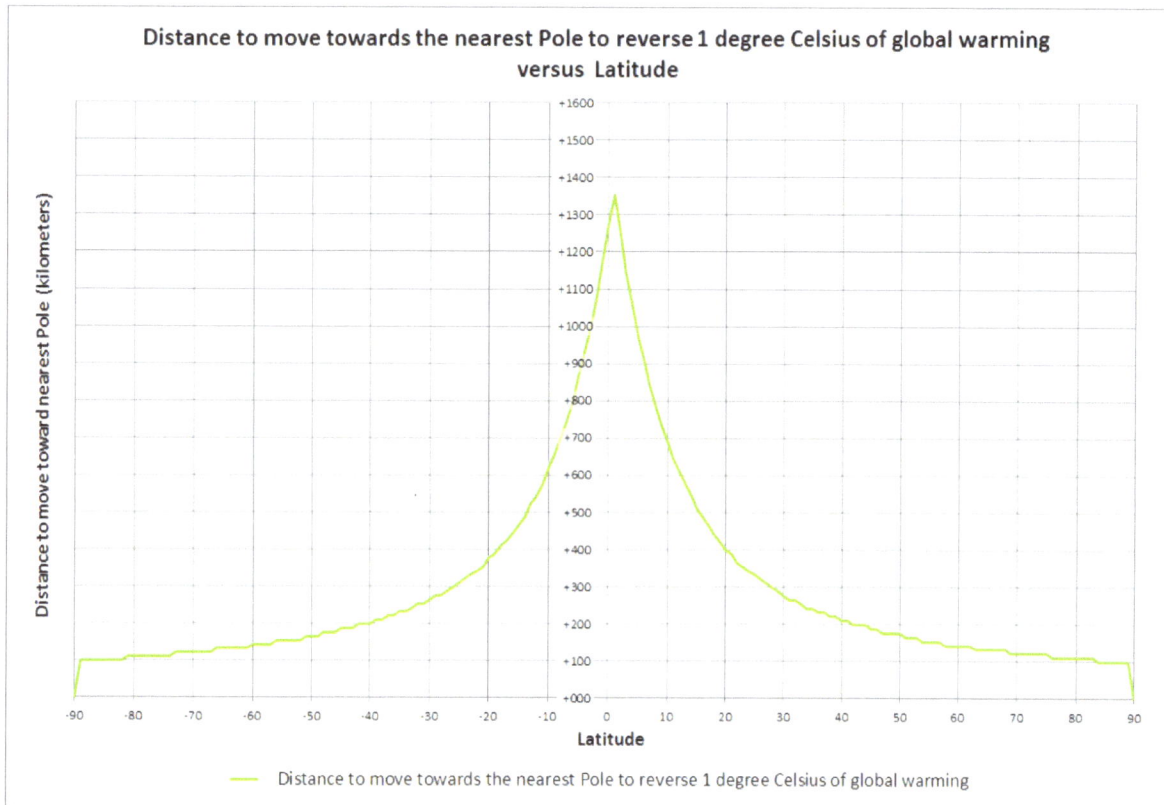

Distance to move towards the nearest Pole to reverse 1 degree Celsius of global warming versus Latitude

An example will show how to use this graph. Imagine that you live in Sydney, in Australia. Your yearly average temperature is about +17 degrees Celsius, and your latitude is about 34 South (or -34). Imagine that global warming raises your yearly average temperature to about +18 degrees Celsius. You decide to move south, to bring your average temperature back to about +17 degrees Celsius. You just need to work out how far to go.

You consult the graph. You find -34 on the X-axis (your latitude), and go vertically up to the green line. When you get to the green line, you go horizontally until you get to the Y-axis. The value on the Y-axis, is the distance that you must drive in the southerly direction, to reverse the 1 degree Celsius of global warming. In this case, you would need to drive a little under 250 km towards the South Pole.

If the distance that you would have to move to reverse 1 degrees Celsius of global warming is not very large, that shows that global warming has not had a very large effect.

But this graph can also be used for something even more interesting. You can use the graph to find out what global warming will be like, before it even happens.

Just work out the distance for your latitude like you did before, but instead of driving towards the nearest Pole, drive towards the Equator (which will increase your average temperature).

When you reach your destination, that is what an additional 1 degree Celsius of global warming will be like, at your original location.

Summary

This book divided the Earth into 14 regions, and then looked at how much warming has occurred in each region. Seven regions had warmed by LESS than the average, and seven regions had warmed by MORE than the average.

Then actual average temperatures, and actual winter temperatures were looked at, which showed that many of the regions which have had the most global warming are the coldest regions. These cold regions are still colder than the regions which have had less global warming. Many of these cold regions will be better with a little global warming.

Global warming was then compared to 3 other examples of warming. These were

  · seasonal warming (winter to summer)
  · the actual temperature in different cities
  · the distance to move to reverse global warming

Seasonal warming was shown to be very fast and large compared to global warming. But people are more worried by very slow and small global warming. This worry is not logical.

The average actual temperature in different cities was shown to vary from cold (Moscow = +4.0 degrees Celsius) to hot (Bangkok = +28 degrees Celsius). Millions of people live in different cities at a huge range of temperatures, with no serious problems. Is it logical to be very worried about another 0.5 or 1.0 degrees Celsius of global warming when you see the huge range of temperatures that people live at?

It was shown that most people can reverse the effect of 1.0 degrees Celsius of global warming by moving 200 to 400 kilometres toward the nearest Pole. The fact that this distance is not large shows that the effect of global warming is not great.

By moving 200 to 400 kilometres towards the Equator, a person can see what an additional 1.0 degrees Celsius of global warming would be like.

The evidence shown in this book makes it clear that global warming will not be as big, or as bad, as many people think.

www.ingramcontent.com/pod-product-compliance
Lightning Source LLC
Chambersburg PA
CBHW050912210326
41597CB00002B/98

9780473559670